Properties of Matter
Atoms, Elements, and Compounds

by Rebecca L. Johnson

Table of Contents

Millmark
EDUCATION

New coins are cut from sheets of **metal**. The coins are bright and shiny. After they are made, the new coins go to banks.

Choose two coins. Describe each of them.

A ____ is ____.

The coin looks ____.

The coin has a picture of ____.

What else do you know about coins? Share your ideas.

metal – a hard, heavy, usually shiny substance

dollar

GEORGE WASHINGTON

1st PRESIDENT 1789–1797

penny

nickel

dime

quarter

half dollar

Coins are cut from sheets of metal like these. The sheets have been rolled up to make them easier to move and handle.

CHAPTER 1

Matter, Matter, Everywhere!

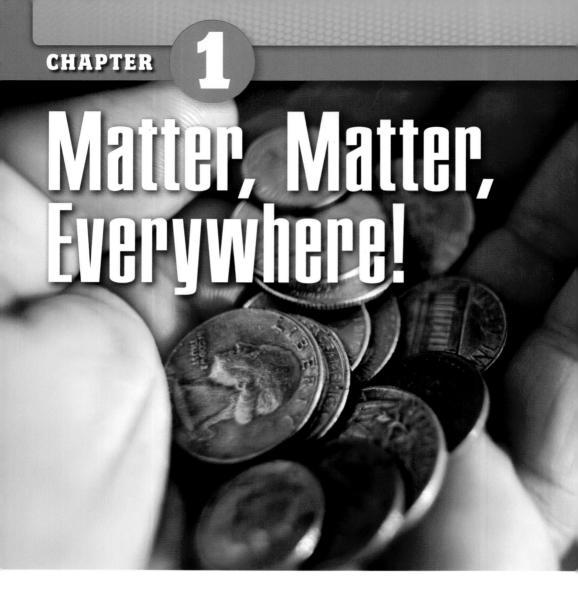

You probably see and touch coins every day. Just a few can fill up your hand. A handful of coins is heavy!

Coins are made of **matter**. Matter is anything that takes up space and has **mass**. You are matter. Everything you see around you is matter, too.

matter – anything that takes up space and has mass

mass – the amount of matter in something

There are many different kinds of matter.
Each kind of matter has different **properties**.
Properties describe what matter is like.

Coins are hard, heavy, and shiny. Coins are also
flat and round. Hard, heavy, shiny, flat, and round
are properties of matter that help describe coins.

properties – qualities of matter that can
be observed or measured

KEY IDEA Properties of
matter help describe
what matter is like.

States of Matter

A very important property of matter is its **state**. Three states of matter are **solid**, **liquid**, and **gas**.

A glass and a pitcher are solids. Solids have a **definite** size and shape.

Orange juice is a liquid. Liquids have a definite size but not a definite shape.

Air is a gas. Gases have no definite shape or size. Gases spread out to fill whatever space they can.

state – the form matter is in

solid – matter that has a definite size and shape

liquid – matter that has a definite size but not a definite shape

gas – matter that has no definite size or shape

definite – clearly defined

◀ The glasses and pitcher are solids. The orange juice is a liquid. The air around the people is a gas.

Changing States

Matter can change from one state into another. For example, ice is frozen water. Ice is a solid. When ice melts, it changes state. It goes from being a solid to being a liquid.

When liquid water boils, some of the water changes state. It goes from being a liquid to being a gas. The gas bubbles to the surface. Then the gas moves into the air.

solid

liquid

gas

KEY IDEAS Solid, liquid, and gas are three states of matter. Matter can change from one state into another.

Measuring Matter

We can **measure** some properties of matter. Length is how long something is. It is often measured in meters. Mass is the amount of matter in something. It is often measured in grams. **Volume** is the amount of space something takes up. It is often measured in liters.

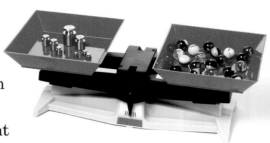

Density is how much mass there is in a certain volume of matter. A liter of pebbles and a liter of water have the same volume. But the liter of pebbles has more mass. So the pebbles have a greater density than the water.

water

pebbles

▲ **The pebbles sink because they have a greater density than water.**

---

measure – find the exact size or amount of something

volume – the amount of space that matter takes up

density – amount of mass in matter with a certain volume

KEY IDEA Some properties of matter can be measured.

INFER

Look at the picture.
Is the ice sinking or floating?

The ice is _____.

What can you infer about the density of water and ice?

The density of _____ is _____ than the density of _____.

MAKE CONNECTIONS

This book is matter. Describe two of its properties that you could measure.

USE THE LANGUAGE OF SCIENCE

What are three states of matter?

Solid, liquid, and gas are three states of matter.

CHAPTER 2

Elements and Compounds

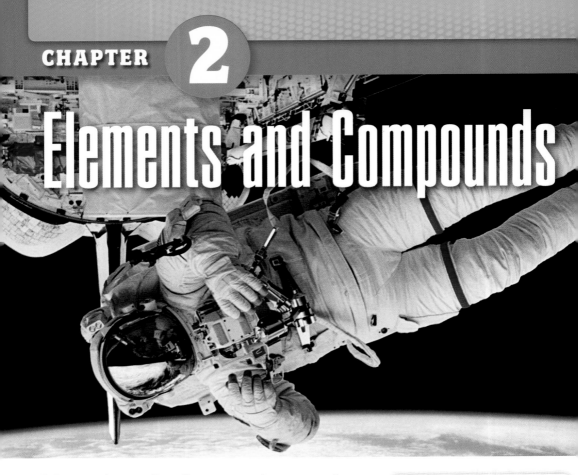

Matter is made of **atoms**. An atom is the smallest whole unit of matter. Most matter is made up of different kinds of atoms. For example, water is made up of two kinds of atoms.

Some matter is made up of only one kind of atom. Gold has only gold atoms. Gold is an **element**. An element is matter made of only one kind of atom.

▲ **The astronaut's suit is made of many different kinds of atoms. The helmet visor is coated with gold.**

gold nugget

atoms – the smallest whole units of matter

element – a type of matter made up of only one kind of atom

Some elements are a solid at room temperature. Copper is an element. It is a solid at room temperature.

Mercury is an element. It is a liquid at room temperature.

Helium is an element. It is a gas at room temperature. Helium has a lower density than air. That is why balloons filled with helium float in air.

KEY IDEA An element is matter made of only one kind of atom.

copper cookware

helium-filled balloons

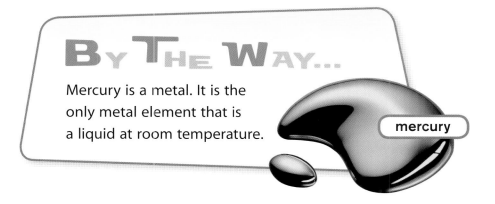

By The Way...

Mercury is a metal. It is the only metal element that is a liquid at room temperature.

mercury

Atoms and Molecules

Some elements are made up of atoms that do not usually **bond**, or join, together with other atoms. The element helium is a good example. In a balloon filled with helium, each helium atom moves around by itself. The helium atoms do not bond together.

helium-filled balloon

Other elements are usually **molecules**. A molecule is two or more atoms joined, or bonded, together. The element oxygen is a gas in the air. Molecules of oxygen are made of two oxygen atoms bonded together.

bond – join very tightly

molecules – sets of two or more atoms that are bonded together

SHARE IDEAS How many atoms are needed to make a molecule? Explain.

two helium atoms

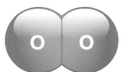

one oxygen molecule

KEY IDEA A molecule is two or more atoms bonded together.

▲ **This drawing shows that two helium atoms do not bond together but two oxygen atoms bond to form a molecule.**

Combining as Compounds

Different elements can bond to form **compounds**. A compound is a molecule that has at least two different elements.

Water is a compound. A molecule of water has one atom of oxygen and two atoms of hydrogen.

Carbon dioxide is another compound. A molecule of carbon dioxide has one atom of carbon and two atoms of oxygen.

compounds – molecules that are made up of two or more elements

Explore Language

Latin Word Roots
compound
com– (with) + *ponere* (to put) = to put with, to put together

▲ **Water has one oxygen atom and two hydrogen atoms.**

▲ **Carbon dioxide has two oxygen atoms and one carbon atom.**

When elements bond together to make compounds, their properties change. The compound they form has new, different properties.

Oxygen and hydrogen are both gases at room temperature. They bond to make liquid water.

Carbon can be in the form of a black solid. When carbon bonds with oxygen, the new compound is a colorless gas.

KEY IDEA A compound is a molecule that has at least two different elements.

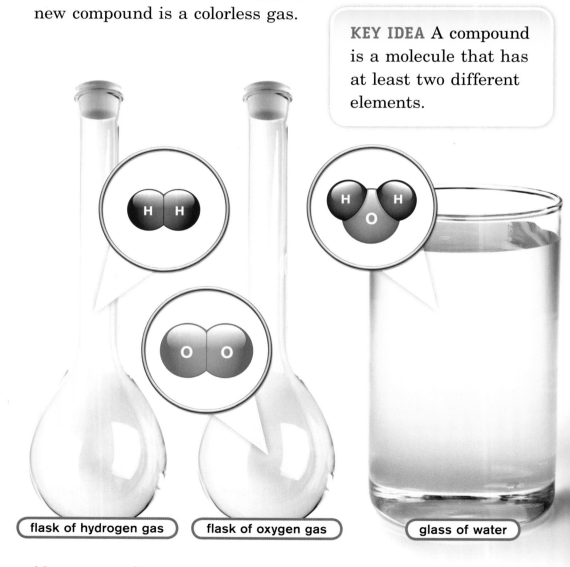

flask of hydrogen gas   flask of oxygen gas   glass of water

COMMUNICATE

This photo shows a model of atoms bonded together. The model communicates information about the atoms. Does the model show an element or a compound? Talk to a friend. Tell how the model helps you answer the question.

MAKE CONNECTIONS

Water is a compound you use and drink every day. Describe three properties of water.

 STRATEGY FOCUS

Make Inferences

Reread pages 10–14. Look at the facts below. Add others. What can you infer from the facts? See the example.

Facts	Inferences
• Helium-filled balloons float in air because helium has a lower density than air. • Copper doesn't float in air.	Copper has a higher density than air.

CHAPTER 3

Organizing the Elements

Thousands of years ago, people knew of just a few elements such as gold and silver. Over time, many more were discovered. Today, scientists know of over 100 different elements.

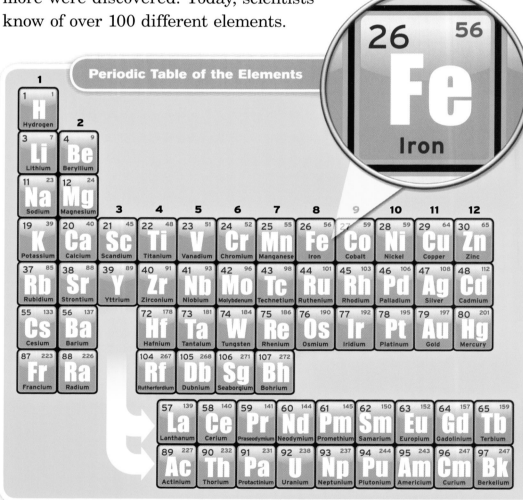

Periodic Table of the Elements

Scientists have organized the elements into a **table**.
The table is called the **Periodic Table of the Elements**.

All of the elements in the periodic table have **symbols**.
The symbols are one or two letters long. Scientists often
use these symbols to talk and write about the elements.

table – a way to organize information

Periodic Table of the Elements – a table
scientists use to organize all the elements

symbols – things that stand for, or represent,
something else

BY THE WAY...

Some element symbols come
from names the elements had
long ago. For example, iron's
symbol is Fe. It comes from the
Latin word for iron, *ferrum*.

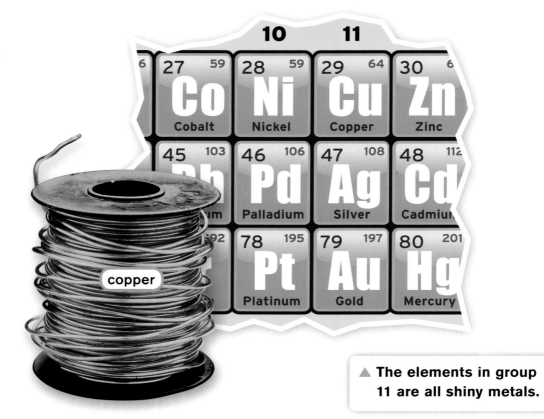

| 6 | 27 59 **Co** Cobalt | 28 59 **Ni** Nickel | 29 64 **Cu** Copper | 30 6 **Zn** Zinc |

| 45 103 **h** m | 46 106 **Pd** Palladium | 47 108 **Ag** Silver | 48 112 **Cd** Cadmiu |

| 92 | 78 195 **Pt** Platinum | 79 197 **Au** Gold | 80 201 **Hg** Mercury |

copper

▲ The elements in group 11 are all shiny metals.

In the periodic table, the elements are arranged into 18 groups. Each column in the table is one group.

Elements in the same group have similar properties. For example, group 11 includes copper, silver, and gold. These elements are all shiny metals. They have other properties in common, too.

SHARE IDEAS Talk about the other properties that the elements in group 11 have in common.

KEY IDEA All the elements are organized into the Periodic Table of the Elements.

INTERPRET DATA

The table below shows the properties of three elements. Which two do you think belong in the same column in the periodic table? Explain.

	Properties		
	Element 1	**Element 2**	**Element 3**
state at room temperature	solid	solid	solid
color	bluish white	pale yellow	grayish white
shininess	shiny	dull	shiny
hardness	hard	crumbles into a powder	fairly hard

MAKE CONNECTIONS

Your bones and teeth contain the element calcium. Find out why calcium is important for your body. Name some foods that contain calcium.

EXPAND VOCABULARY

Compound refers to a combination of two or more things.

Look at these sentences:

Water is a **compound** of hydrogen and oxygen.

Baseball is a **compound** word.

They live in a four-building **compound**.

Compare the meaning of **compound** in all the sentences.

Write and draw pictures to show the similarities.

What Is a Silversmith?

A silversmith makes jewelry and other objects out of silver. Silversmiths may also work with other elements, such as gold and copper.

Read the chart to see if you would like to be a silversmith.

Would I like this career?	**You might like this career if:** • you are artistic. • you like making things with your hands.
What would I do?	• You would work with silver and other metals. • You might make jewelry or repair antique metal objects.
How can I prepare?	• Take art classes. • Visit with a silversmith and learn about what he or she does.

Use *What* and *How*

What and **how** are used to ask questions. **What** can be used to ask for a definition. You can use part of the question in the answer.

EXAMPLE

What is matter?

Matter is anything that takes up space and has mass.

How is used to ask about the way that something happens.

EXAMPLE

How are water and ice different?

Water is a liquid and ice is a solid.

With a friend, ask each other questions about elements and compounds. Use **what** and **how**. Answer in complete sentences.

Write Questions and Answers

You learned that solids, liquids, and gases are states of matter. Choose a liquid in your home. Ask questions about the different states of this liquid. Research the answers.

- Use **what** and **how**.
- Ask questions and give the answers.
- Use words from the questions in your answers.

Words You Can Use
What is…
What are…
How is…
How are…

Elements in a Bottle

Some foods contain a lot of vitamins and minerals. Many vitamin pills also contain vitamins and minerals. Like vitamins, minerals are substances people need to live and grow. Many minerals are elements.

Look at the label in the picture. What minerals do you see?

Turn to the Periodic Table of the Elements on pages 16 and 17. Find these elements in the table.

What are the symbols for the elements?

Each Tablet Contains	%DV
Magnesium 100 mg	25%
Zinc 15 mg	100%
Selenium 20 mcg	29%
Copper 2 mg	100%
Manganese 2 mg	100%
Chromium 120 mcg	100%
Molybdenum 75 mcg	100%
Chloride 72 mg	2%
Potassium 80 mg	2%
Boron 150 mcg	*

• Look at the larger view of the label. These substances are minerals. Name the minerals you see.

• Turn to the Periodic Table of the Elements on pages 16-17. Which of these minerals are elements on the table?

• What are the symbols for these elements?

Key Words

atom (atoms) the smallest whole unit of matter
An **atom** is too small to see.

bond join very tightly
Oxygen and hydrogen atoms can **bond** together to form water.

compound (compounds) a molecule that is made up of two or more elements
Water is a **compound**.

element (elements) matter made of only one kind of atom
Gold is an **element**.

gas (gases) matter that has no definite size or shape
Oxygen is a **gas** found in air.

liquid (liquids) matter that has a definite size but not a definite shape
A **liquid** takes the shape of its container.

mass (masses) the amount of matter in something
A balance measures an object's **mass**.

matter anything that takes up space and has mass
Everything around you is made of **matter**.

molecule (molecules) two or more atoms bonded together
A **molecule** of oxygen has two atoms.

Periodic Table of the Elements a table scientists use to organize all the elements
Each element has its own symbol in the **Periodic Table of the Elements**.

property (properties) a quality of matter that can be observed or measured
Volume is a **property** that can be measured.

solid (solids) matter that has both a definite size and shape
A rock is a **solid**.

state (states) the form matter is in
Ice is the solid **state** of water.

Index

MILLMARK EDUCATION CORPORATION
Ericka Markman, President and CEO; Karen Peratt, VP, Editorial Director; Lisa Bingen, VP, Marketing; David Willette, VP, Sales; Rachel L. Moir, VP, Operations and Production; Shelby Alinsky, Editor; Mary Ann Mortellaro, Science Editor; Amy Sarver, Series Editor; Betsy Carpenter, Editor; Guadalupe Lopez, Writer; Kris Hanneman and Pictures Unlimited, Photo Research

PROGRAM AUTHORS
Mary Hawley; Program Author, Instructional Design
Kate Boehm Jerome; Program Author, Science

BOOK DESIGN Steve Curtis Design

CONTENT REVIEWER
Carla C. Johnson, EdD, University of Toledo, Toledo, OH

PROGRAM ADVISORS
Scott K. Baker, PhD, Pacific Institutes for Research, Eugene, OR
Carla C. Johnson, EdD, University of Toledo, Toledo, OH
Margit McGuire, PhD, Seattle University, Seattle, WA
Donna Ogle, EdD, National-Louis University, Chicago, IL
Betty Ansin Smallwood, PhD, Center for Applied Linguistics, Washington, DC
Gail Thompson, PhD, Claremont Graduate University, Claremont, CA
Emma Violand-Sánchez, EdD, Arlington Public Schools, Arlington, VA (retired)

TECHNOLOGY
Arleen Nakama, Project Manager
Audio CDs: Heartworks International, Inc.
CD-ROMs: Cannery Agency
ResourceLinks Website: Six Red Marbles

PHOTO CREDITS Cover © Nicholas Pitt/Alamy; 1 © ukrphoto/Shutterstock; 2a, 3a, 3b, 3c, 3d, 3e United States Mint Image; 2b © AP Images/The Denver Post/Glenn Asakawa; 4 © Bob Torrez/PhotoEdit; 5a © Photodisc Green/Punchstock; 5b © Joellen L. Armstrong/Shutterstock; 6 © Myrleen Ferguson Cate/PhotoEdit; 7a © Adam Hart-Davis/Photo Researchers, Inc.; 7b and 18a © Charles D. Winters/Photo Researchers, Inc.; 8a, 14c, 22 Ken Cavanagh for Millmark Education; 8b © Navaswan/Getty Images; 9a © FoodPix/Jupiter Images; 9b and 9c Lloyd Wolf for Millmark Education; 10a © NASA Johnson Space Center (NASA-JSC); 10b © Steffen Foerster Photography/Shutterstock; 11a © Photodisc/Punchstock; 11b © Stephanie Rausser/Getty Images; 11c © ImageState/Alamy; 12 © S.K.I.L. National Helium Services; 13 © Tihis/Shutterstock; 14a © Mike Kemp/Rubberball Productions/Getty Images; 14b © Siede Preis/Getty Images; 15 © Blend Images / Alamy; 16–17 and 18b illustrations by Steve Curtis Design; 20a © David Gordon/Alamy; 20a © David Gordon/Alamy; 23 © Tjerrie Smit/Shutterstock; 24 © Joshua Haviv/Shutterstock

Copyright © 2008 Millmark Education Corporation

All rights reserved. Reproduction of the whole or any part of the contents without written permission from the publisher is prohibited. Millmark Education and ConceptLinks are registered trademarks of Millmark Education Corporation.

Published by Millmark Education Corporation
PO Box 30239
Bethesda, MD 20824

ISBN-13: 978-1-4334-0057-5

Printed in the USA

10 9 8 7 6 5 4 3